EMMANUEL JOSEPH

Brushstrokes of the Earth, Art, Psychology, and the Fight for Environmental Healing

Contents

1

Chapter 1: The Canvas of Nature

Nature is a master artist, and its canvas stretches across the globe in every direction. From the vibrant hues of a sunset to the intricate designs on a butterfly's wings, the natural world is a living gallery. The interconnectedness of all living things forms a vast, complex tapestry, with each element contributing to the overall picture. This chapter delves into the awe-inspiring beauty of nature, exploring how its artistic qualities have inspired humans for centuries. Through vivid descriptions and personal anecdotes, we'll journey through lush forests, serene lakes, and towering mountains, capturing the essence of nature's artistry.

Imagine standing at the edge of a dense forest, the air thick with the scent of pine and the sound of rustling leaves. As the sunlight filters through the canopy, it creates a kaleidoscope of colors on the forest floor. This symphony of light and shadow is just one example of how nature uses its brushstrokes to create a masterpiece. In the heart of the Amazon rainforest, the vibrant colors of exotic birds and flowers add to the richness of the landscape, creating a living mosaic that is both beautiful and fragile.

The serene lakes of the world offer another perspective on nature's artistry. Picture a calm, crystal-clear lake reflecting the sky above, its surface undisturbed except for the occasional ripple from a fish or the breeze. The tranquility of these waters provides a sense of peace and stillness, inviting us to pause and appreciate the beauty around us. In Japan, the famous cherry blossom season transforms the landscape into a sea of pink and white, a fleeting yet breathtaking display of nature's ability to captivate and inspire.

Towering mountains, with their majestic peaks and rugged terrain, offer a different kind of beauty. The sheer scale and grandeur of these natural wonders remind us of the power and resilience of the Earth. Whether it's the snow-capped peaks of the Himalayas or the ancient volcanoes of the Andes, these mountains stand as testament to the enduring strength of nature. As we explore these varied landscapes, we begin to see how each element of nature contributes to the larger canvas, creating a dynamic and ever-changing work of art.

2

Chapter 2: The Mind's Palette

The human mind is an artist's palette, filled with colors that represent our emotions, thoughts, and experiences. Just as an artist blends colors to create a masterpiece, our minds combine various elements to shape our perception of the world. This chapter delves into the psychological aspects of creativity, examining how art can be a therapeutic tool for healing and self-expression. We'll explore the stories of individuals who have used art to overcome trauma and find solace, highlighting the transformative power of creative expression.

Consider the story of Sarah, a war veteran who found herself struggling with post-traumatic stress disorder (PTSD) after returning home. Traditional therapy didn't seem to help, and Sarah felt trapped by her memories. It was only when she picked up a paintbrush and began to translate her emotions onto canvas that she found a sense of relief. The act of painting allowed her to externalize her pain and start the healing process. Sarah's journey illustrates how art can serve as a powerful outlet for processing complex emotions.

In another instance, a community in the aftermath of a natural disaster came together to create a mural that depicted their shared experiences and hopes for the future. This collaborative project not only beautified their environment but also provided a sense of unity and healing. The mural became a symbol of resilience and collective strength, demonstrating how art can bring people together and foster a sense of community.

Art therapy is a growing field that harnesses the therapeutic potential of creative expression. Through activities such as drawing, painting, and sculpting, individuals are encouraged to explore their inner worlds and express their emotions in a safe and supportive environment. This process can lead to increased self-awareness, emotional regulation, and personal growth. As we delve into the stories of those who have found healing through art, we begin to understand the profound impact that creative expression can have on the human psyche.

3

Chapter 3: Environmental Crisis as an Artistic Challenge

The environmental crisis is not just a scientific or political issue; it's also an artistic challenge. This chapter examines how artists around the world are using their craft to raise awareness about environmental issues and inspire change. From sculptures made of recycled materials to powerful murals depicting climate change, art has become a vital tool in the fight for environmental healing. We'll delve into the works of renowned eco-artists and uncover the stories behind their creations, showcasing how art can be a catalyst for environmental advocacy.

One notable example is the work of artist John Dahlsen, who creates stunning sculptures and installations using debris collected from beaches. His art not only highlights the issue of ocean pollution but also transforms waste into something beautiful, challenging viewers to reconsider their relationship with consumption and waste. Dahlsen's work reminds us that even the most discarded materials can be repurposed and given new life.

In the streets of cities worldwide, muralists are using their talents to spread messages of environmental consciousness. A striking mural in London, created by artist Louis Masai, depicts a melting ice cream cone with endangered animals, symbolizing the rapid decline of biodiversity due to climate change. This powerful visual serves as a call to action, urging passersby to reflect on their impact on the planet and take steps toward sustainability.

Artists are also exploring new mediums to address environmental issues. Digital artist Joanie Lemercier uses light projections and mapping to create immersive experiences that draw attention to the fragility of our ecosystems. Her installations transport viewers to environments at risk of disappearing, fostering a deep emotional connection to the natural world. These innovative approaches demonstrate how art can evolve to meet the challenges of our time and inspire a collective effort to protect the planet.

4

Chapter 4: The Psychology of Environmentalism

Understanding the psychological motivations behind environmentalism is crucial for fostering a sustainable future. This chapter explores the psychological theories that explain why people care about the environment and how these motivations can be harnessed to promote pro-environmental behavior. We'll delve into concepts such as eco-anxiety and biophilia, and share stories of individuals who have made significant lifestyle changes to reduce their environmental impact. Through these narratives, we'll gain insights into the human psyche and its relationship with the natural world.

Eco-anxiety, a term that has gained prominence in recent years, refers to the feelings of fear and helplessness that arise from concerns about environmental degradation. While these emotions can be overwhelming, they can also serve as a powerful motivator for action. Take the story of Emma, a young activist who channeled her eco-anxiety into organizing community clean-up events and advocating for policy changes. Her journey illustrates how confronting eco-anxiety can lead to meaningful contributions toward environmental preservation.

Biophilia, the innate human affinity for nature, is another key concept in understanding environmentalism. This deep-seated connection to the natural world influences our desire to protect and preserve it. Consider the story of Sam, a city dweller who found solace in tending to an urban garden. This small act of nurturing plants not only brought Sam joy but also fostered a sense of responsibility toward the environment. Sam's experience highlights how biophilia can inspire everyday actions that contribute to environmental sustainability.

Psychologists have also identified the role of social norms in shaping pro-environmental behavior. When individuals see their peers engaging in eco-friendly practices, they are more likely to adopt similar behaviors. This phenomenon can be observed in the rise of zero-waste communities, where members collectively strive to minimize their environmental footprint. By understanding the psychological drivers of environmentalism, we can develop strategies to encourage widespread adoption of sustainable practices.

5

Chapter 5: Art as a Reflection of Environmental Consciousness

Art has always been a mirror reflecting the values and concerns of society. In this chapter, we'll examine how contemporary art reflects our growing environmental consciousness. From the resurgence of land art to the rise of digital art, artists are finding innovative ways to address environmental themes. We'll explore exhibitions and installations that challenge viewers to rethink their relationship with nature, and share stories of artists who are pushing the boundaries of traditional art forms to make powerful environmental statements.

The land art movement, which emerged in the 1960s, continues to inspire contemporary artists who use the natural landscape as both canvas and medium. One notable example is the work of artist Andy Goldsworthy, who creates ephemeral sculptures using materials found in nature. His installations, such as intricate stone spirals or ice sculptures that melt with the rising sun, invite viewers to contemplate the transient beauty of the natural world and the impact of human activity on it.

Digital art has opened up new possibilities for addressing environmental themes. Artists like Refik Anadol use data-driven installations to visualize the effects of climate change, creating immersive experiences that make abstract concepts tangible. Anadol's work, which often incorporates real-time data, allows viewers to witness the ongoing changes in our environment and reflect on their implications.

Exhibitions dedicated to environmental art provide a platform for artists to engage with audiences on critical issues. The "Earth Matters" exhibition at the National Museum of African Art showcased works by African artists addressing topics such as deforestation, pollution, and water scarcity. These powerful pieces not only highlight the environmental challenges facing the continent but also celebrate the resilience and creativity of its people. Through these exhibitions, art serves as a conduit for dialogue and action, fostering a deeper understanding of our environmental responsibilities.

6

Chapter 6: Nature's Influence on Artistic Expression

ature has long been a source of inspiration for artists, influencing their work in profound ways. This chapter delves into the historical and cultural significance of nature in art, from ancient cave paintings to modern-day masterpieces. We'll explore how different cultures interpret nature through their art and share stories of artists who have drawn inspiration from the natural world. Through these narratives, we'll gain a deeper appreciation for the enduring connection between art and nature.

Ancient cave paintings, such as those found in Lascaux, France, provide some of the earliest evidence of humans depicting the natural world. These vivid portrayals of animals and

Ancient cave paintings, such as those found in Lascaux, France, provide some of the earliest evidence of humans depicting the natural world. These vivid portrayals of animals and scenes from daily life offer a glimpse into the relationship between early humans and their environment. The meticulous detail and reverence for nature evident in these works reflect a deep respect for the natural world that has persisted throughout history.

Different cultures have interpreted nature in unique ways through their art. For example, traditional Japanese ink wash paintings, known as sumie, capture the essence of landscapes and natural forms with minimalistic brushstrokes. These works emphasize the beauty of simplicity and the harmony between humans and nature. Similarly, Indigenous Australian dot paintings use intricate patterns to represent the natural world and convey cultural stories and knowledge passed down through generations.

In the realm of modern art, the influence of nature can be seen in the works of artists like Georgia O'Keeffe, who is renowned for her large-scale paintings of flowers and desert landscapes. O'Keeffe's art captures the delicate beauty and vibrant colors of nature, inviting viewers to appreciate the intricacies of the natural world. Her work reflects a personal connection to the environment and a desire to share that connection with others.

Through these diverse expressions, we see how nature has inspired artists across cultures and time periods. The enduring connection between art and nature reminds us of our shared responsibility to protect and preserve the environment for future generations.

7

Chapter 7: The Role of Art in Environmental Education

Art has the power to educate and inspire, making it an effective tool for environmental education. This chapter explores how art is being used in schools, communities, and public spaces to teach people about environmental issues and encourage sustainable practices. We'll share stories of innovative art-based education programs and highlight the work of artists who are collaborating with educators to create impactful learning experiences. Through these examples, we'll see how art can be a bridge between knowledge and action.

One innovative program is the "Eco-Art" initiative, which partners artists with schools to create environmental art projects. Students are encouraged to explore environmental themes through creative expression, such as making sculptures from recycled materials or painting murals that depict local wildlife. These projects not only foster creativity but also raise awareness about environmental issues and promote a sense of stewardship among young people.

Public art installations also play a significant role in environmental education. In the city of Melbourne, the "Urban Forest" project transformed a vacant lot into a community garden and art space. Local artists created sculptures and murals that highlight the importance of urban green spaces and biodiversity. The project engages residents in hands-on activities, such as planting trees and maintaining the garden, fostering a connection to nature and a commitment to environmental sustainability.

Artists are also collaborating with environmental organizations to create impactful educational campaigns. For example, the "Plastic Ocean" project, led by artist Mandy Barker, uses striking photographic images of ocean plastic pollution to raise awareness about the issue. Barker's work has been exhibited in galleries and museums worldwide, sparking conversations about the impact of plastic waste on marine ecosystems and inspiring action to reduce plastic consumption.

Through these examples, we see how art can be a powerful tool for environmental education, bridging the gap between knowledge and action. By engaging people in creative experiences, art fosters a deeper understanding of environmental issues and motivates individuals to make positive changes in their lives.

8

Chapter 8: Healing Through Artistic Creation

Creating art can be a powerful form of healing, both for individuals and communities. This chapter delves into the therapeutic benefits of art, exploring how artistic expression can help people cope with environmental loss and trauma. We'll share stories of communities that have used art to rebuild and recover after natural disasters, and highlight the work of art therapists who are helping individuals process their emotions through creative practices. Through these narratives, we'll understand the healing power of art in the face of environmental challenges.

In the aftermath of Hurricane Katrina, the city of New Orleans faced immense destruction and loss. To help residents cope with the trauma and rebuild their sense of community, local artists organized the "Healing Arts" initiative. This project invited survivors to participate in creating murals, sculptures, and performances that expressed their experiences and hopes for the future. The collective act of creating art provided a sense of solidarity and empowerment, helping the community to heal and rebuild.

Art therapists play a crucial role in helping individuals process their emotions through creative practices. Consider the story of Alex, who struggled with feelings of hopelessness and despair after witnessing the devastation of a wildfire in their hometown. Through art therapy sessions,

Alex was encouraged to use painting and drawing to express their grief and find a sense of closure. The process of creating art allowed Alex to externalize their emotions and gradually come to terms with the loss.

Communities affected by environmental disasters often turn to art as a means of recovery and resilience. In the Philippines, after Typhoon Haiyan, local artists collaborated with residents to create a series of murals that depicted their journey of rebuilding and recovery. These vibrant works of art not only beautified the landscape but also served as a testament to the community's strength and determination. The murals became a source of pride and hope, reminding residents of their capacity to overcome adversity.

Through these stories, we see how art can be a powerful tool for healing in the face of environmental challenges. By providing an outlet for expression and fostering a sense of community, artistic creation helps individuals and communities navigate the complexities of loss and trauma.

9

Chapter 9: The Intersection of Science and Art

The worlds of science and art may seem disparate, but they are deeply interconnected. This chapter explores the ways in which artists and scientists are collaborating to address environmental issues. We'll share stories of interdisciplinary projects that blend scientific research with artistic expression, creating powerful works that communicate complex environmental concepts in accessible ways. Through these examples, we'll see how the fusion of science and art can lead to innovative solutions and inspire collective action.

One notable example is the "Data Art" project, which brings together scientists and artists to visualize environmental data through creative mediums. By transforming raw data into compelling visual narratives, these projects make complex scientific concepts more accessible to the public. For instance, artist Giorgia Lupi's work visualizes data on air pollution in urban areas, using intricate patterns and colors to convey the severity of the issue. Her art not only informs but also evokes an emotional response, prompting viewers to consider their role in addressing air quality.

Another example is the collaboration between marine biologists and sculptor Jason deCaires Taylor. Taylor's underwater sculptures, installed off the coast of various locations, serve as artificial reefs that promote marine biodiversity. These sculptures not only provide a habitat for marine life but also raise awareness about the importance of coral reefs and the threats they face from climate change and pollution. By merging art and science, Taylor's work creates a tangible connection between humans and the underwater world.

Interdisciplinary projects also extend to educational initiatives. The "Science Through Art" program partners artists with scientists to develop interactive exhibits that teach visitors about environmental issues. These exhibits use a combination of art, technology, and hands-on activities to engage audiences of all ages. One such exhibit, focused on climate change, allows visitors to manipulate virtual environments and observe the effects of different actions on the planet's health. By blending science and art, these initiatives foster a deeper understanding of environmental challenges and inspire action.

Through these examples, we see how the intersection of science and art can lead to innovative solutions and inspire collective action. By harnessing the strengths of both disciplines, interdisciplinary projects create powerful works that communicate environmental concepts and motivate individuals to make a positive impact.

10

Chapter 10: Art as a Catalyst for Social Change

Art has the power to inspire social change, and this chapter examines how artists are using their work to advocate for environmental justice. We'll explore the stories of activist artists who are addressing issues such as pollution, deforestation, and climate change through their art. By shining a light on these issues, artists are galvanizing communities and sparking conversations that lead to real-world change. Through these narratives, we'll understand the role of art as a catalyst for environmental activism.

Activist artist Olafur Eliasson is known for his large-scale installations that address environmental issues. His "Ice Watch" project, which involved placing blocks of melting ice in public spaces, aimed to raise awareness about the rapid melting of Arctic ice due to climate change. The melting ice served as a poignant reminder of the urgent need for action, prompting viewers to consider their role in addressing the climate crisis.

In the Amazon rainforest, artist Ernesto Neto has created immersive installations that highlight the importance of preserving indigenous cultures and the natural environment. His works, which often incorporate materials such as spices, seeds, and fibers, invite viewers to engage with the sensory qualities of the rainforest. Neto's art raises awareness about deforestation and the impact of industrial activities on indigenous communities, encouraging viewers to support conservation efforts.

Artists are also addressing issues of environmental justice in urban areas. In Flint, Michigan, artist Hubert Massey created a mural that depicts the city's struggle with lead-contaminated water and the resilience of its residents. The mural serves as a powerful symbol of the community's fight for clean water and environmental justice. By capturing the stories and experiences of Flint residents, Massey's art amplifies their voices and draws attention to the need for systemic change.

Through these stories, we see how art can be a catalyst for social change, shining a light on critical environmental issues and inspiring collective action. By using their talents to advocate for environmental justice, artists play a vital role in galvanizing communities and sparking conversations that lead to real-world change.

11

Chapter 11: The Future of Environmental Art

As we look to the future, the role of art in environmental advocacy will continue to evolve. This chapter explores emerging trends in environmental art and predicts how artists will respond to the ongoing environmental crisis. We'll delve into the possibilities of new technologies, such as virtual reality and augmented reality, in creating immersive art experiences that raise awareness about environmental issues. Through these speculations, we'll imagine the future of environmental art and its potential to inspire a sustainable world.

Virtual reality (VR) and augmented reality (AR) are poised to revolutionize the way we engage with environmental art. These technologies allow artists to create immersive experiences that transport viewers to different environments, providing a deeper understanding of the issues at hand. For instance, a VR installation could take viewers on a journey through a rainforest, highlighting the impact of deforestation, or immerse them in an underwater world affected by plastic pollution. By making these experiences interactive and emotionally engaging, artists can foster a stronger connection to the natural world and inspire action.

Emerging trends also include the use of bio-art, which incorporates living organisms into artistic creations. Artists are exploring the possibilities of working with plants, bacteria, and other biological materials to create dynamic, ever-changing artworks. These pieces not only challenge our perception of art but also emphasize the interconnectedness of all living things. For example, artist Heather Ackroyd's grass installations use the natural growth process of grass to create large-scale portraits and landscapes. These living artworks remind viewers of the beauty and fragility of life and the importance of environmental stewardship.

Another exciting development is the rise of eco-friendly art practices. Artists are increasingly mindful of the environmental impact of their materials and processes, opting for sustainable alternatives. This shift is leading to the creation of artworks that not only convey environmental messages but also embody sustainable principles. For example, artists are using natural pigments, recycled materials, and low-energy production methods to minimize their ecological footprint. This trend highlights the role of artists as leaders in promoting sustainable practices within the creative community.

As we look ahead, the future of environmental art holds immense potential to drive positive change. By embracing new technologies, bio-art, and sustainable practices, artists can continue to inspire and mobilize individuals to take action for the planet. Through these innovations, art will remain a powerful force in the fight for environmental healing.

12

Chapter 12: Personal Reflections and Call to Action

In the final chapter, we reflect on the journey we've taken through the pages of this book and consider the ways in which art and psychology intersect with environmental healing. We'll share personal stories and reflections, encouraging readers to find their own creative paths to environmental advocacy. Through these narratives, we'll emphasize the importance of individual action and collective effort in the fight for a healthier planet. This chapter serves as a call to action, inspiring readers to use their creativity and passion to make a positive impact on the world.

Reflecting on the diverse stories and examples shared throughout this book, it is evident that art has the power to transcend boundaries and connect us to the natural world in profound ways. Whether through a painting, a sculpture, or a performance, art has the ability to evoke emotions, spark conversations, and inspire action. It reminds us of the beauty and fragility of our planet and calls us to be better stewards of the environment.

Consider the story of Maya, an artist who transformed her passion for painting into a platform for environmental advocacy. Maya began creating a series of paintings that depicted endangered species in their natural habitats, highlighting the urgent need for conservation. Her work caught the attention of local environmental organizations, and she started collaborating with them to raise awareness and funds for wildlife protection. Maya's journey illustrates how one individual's creative efforts can make a meaningful difference.

Each of us has the potential to contribute to environmental healing in our own unique ways. Whether you are an artist, a scientist, an educator, or simply someone who cares about the planet, your actions matter. By finding creative ways to engage with environmental issues, we can collectively build a more sustainable and resilient world. This could mean participating in community art projects, supporting eco-friendly initiatives, or using your talents to educate and inspire others.

As we conclude this journey, let us remember that the fight for environmental healing is not just a task for a few—it is a shared responsibility that requires the collective effort of all. By harnessing the power of art and psychology, we can create a future where the beauty of nature is preserved and celebrated for generations to come.

Book Title: Brushstrokes of the Earth: Art, Psychology, and the Fight for Environmental Healing

Description: "Brushstrokes of the Earth" is an illuminating journey into the intersection of art, psychology, and environmentalism. This captivating book explores how the beauty and complexity of the natural world have inspired artistic expression throughout history, and how contemporary artists are using their work to raise awareness about environmental issues and drive meaningful change.

Through twelve engaging chapters, readers will uncover the therapeutic power of art, delve into the psychological motivations behind environmentalism, and discover innovative art projects that blend scientific research with creative expression. From personal stories of healing through artistic creation to the role of art in environmental education, this book provides a comprehensive and inspiring look at how art can be a powerful catalyst for environmental advocacy.

With vivid descriptions, compelling narratives, and thought-provoking insights, "Brushstrokes of the Earth" invites readers to appreciate the profound connection between art and nature. It emphasizes the importance of individual action and collective effort in the fight for a sustainable future, encouraging readers to use their creativity and passion to make a positive impact on the world.

Perfect for artists, environmentalists, psychologists, and anyone interested in the transformative power of creativity, this book is a celebration of the enduring bond between humanity and the natural world.